A Cumbria G

THE APPLES AND ORCHARDS OF CUMBRIA

A comprehensive review of Cumbrian apple varieties
and of Cumbrian orchards open to the public

ANDY GILCHRIST

HAYLOFT PUBLISHING LTD
CUMBRIA

First published by Hayloft Publishing Ltd., 2013

Hayloft Publishing Ltd, South Stainmore,
Kirkby Stephen, Cumbria, CA17 4DJ

tel: 017683 41568 or 07971 352473
email: books@hayloft.eu
web: www.hayloft.eu

Copyright © Andy Gilchrist, 2013

ISBN 978 190 452 4557

A CIP catalogue record for this book is available from the British Library

Apart from any fair dealing for the purposes of research or private study or criticism or review, as permitted under the Copyright Designs and Patents Act 1988 this publication may only be reproduced, stored or transmitted in any form or by any means with the prior permission in writing of the publishers, or in the case of reprographic reproduction in accordance with the terms of the licenses issued by the Copyright Licensing Agency.

Designed, printed and bound in the EU

Papers used by Hayloft are natural, recyclable products made from wood grown in sustainable forests. The manufacturing processes conform to the environmental regulations of the country of origin.

This book is dedicated to Hilary Wilson, Cumbria's own 'Queen of Apples', without whose research and grafting work, some of the rediscovered Cumbrian varieties described herein might have been lost forever

THE APPLES AND ORCHARDS OF CUMBRIA

Contents

INTRODUCTION AND HISTORY… 7
INTRODUCTION TO CUMBRIAN APPLE VARIETIES 10
CUMBRIAN APPLE VARIETY DESCRIPTIONS: 13
    Autumn Harvest,
    Bradley's Beauty,
    Carlisle Codlin,
    Churn Lid,
    Duke of Devonshire,
    Fallbarrow Favourite,
    Forty Shilling,
    Greenup's Pippin,
    John Hugget,
    Keswick Codlin,
    Lancashire Pippin,
    Lemon Square,
    Longstart,
    Nelson's Favourite,
    Rankthorn,
    Royal,
    Taylor's Favourite,
    Tiffen,
    Wheaten Loaves
    Varieties originating elsewhere but widely grown in Cumbria…
    Lost Varieties and Seedling Varieties…
CUMBRIAN ORCHARDS TO VISIT…… 29
MAP OF CUMBRIA SHOWING ORCHARD LOCATIONS……… 30
    Acorn Bank,
    Ashmeadow,
    Brantwood,
    Cylinders Estate,
    Dalemain,
    Grange-over-Sands Community Orchard,

    Hutton-in-the-Forest,
    Lakeside Hotel,
    Leighton Hall,
    Levens Hall,
    Mirehouse,
    Rydal Hall,
    Sizergh Castle,
    Winderwath Gardens,
    Wordsworth House and Garden

SMALLER ORCHARDS: 71
    Brockhole,
    Cowgill Community Orchard,
    Growing Well,
    Hallgarth Community Orchard,
    SLOG Orchard,
    NGS gardens with fruit

OTHER TREE FRUIT GROWN IN CUMBRIA…   76
    Pear,
    Damson,
    Plum,
    Cherry,
    Quince,
    Medlar,
    Mulberry

FUTURE PROSPECTS FOR APPLES & ORCHARDS IN CUMBRIA…   80
ACKNOWLEDGEMENTS…   83

Introduction and History

THIS BOOK SETS out to describe, in two parts, firstly the apple varieties which originated in the region now known as Cumbria, and secondly the Cumbrian gardens and/or orchards open to the public where those and other varieties may be seen and enjoyed. In the world of fruit-growing, there is arguably no finer sight than an orchard in full blossom, unless it be that same orchard heaving with ripe fruit at harvest time.

Ask most people where apples are grown in England and, if they are able to answer at all, they may suggest Kent or Somerset. Few if any would say Cumbria because there is no significant commercial production of apples here today. Yet it was not always thus as place names attest: Appleby (a farm or settlement where apple trees were grown), Applethwaite (a forest clearing where apple trees were grown), Plumgarth (an enclosure where plum trees were grown) etc. Of course apples did not originate here or even in Europe.

Apples are the fruit of *Malus domestica*, a species which has several thousand cultivars worldwide. The genus (mainly *Malus sieversii*, *Malus orientalis* and *Malus sylvestris*) originated in the Tien Shan mountains of Kazakhstan, where wild apple forests still exist today. It is no coincidence that Kazakhstan's former capital was named Alma Ata or 'Father of Apples.'

Domestication of apple trees has been dated as far back as 6,500BC in the Middle East but the most significant breakthrough in apple cultivation was the development of the science or art of grafting onto selected rootstocks in approx 2,500BC. This permitted the propagation and continued cultivation of genetically identical trees of any desired variety. The basic principles of grafting have not changed significantly to the present day, although more trees are now propagated commercially by chip budding than grafting.

Successive civilisations such as Assyrians, Persians, Greeks and

Romans all valued apples highly both for the beauty of the (usually walled) orchard which provided blossom, fragrance and shade, and for the decorative and culinary value of the fruit itself. The earliest reference to named varieties can be found in the writing of the Roman historian Pliny who listed over twenty apple varieties in his *Natural History* in the first century AD. Fruit trees even had their own deity: the goddess Pomona.

It is believed the Romans first brought the domestic apple to Britain, as opposed to the wild crab apple, *Malus sylvestris*, which had been resident here for millenia. When the Romans left, cultivated apples were maintained through the dark ages by monks in the Abbeys, who valued apples not only for culinary uses but also for juice and cider.

Following the Norman invasion, new European apple varieties were brought to England and orchard plantings increased all over the country on the estates of the rural barons and aristocracy. Here in Lakeland two distinct categories of fruit growing developed.

The great country houses grew a wide variety of fruit, often in walled gardens. Apples were eaten fresh at the conclusion of a meal, with varieties ranging from earlies ripening in late August through to late varieties which could be stored into the New Year in specially built fruit stores, providing a continuous supply of fresh apples for half the year. The gardeners of many country houses raised new varieties which would typically be named in honour of the aristocracy, thus the many varieties entitled *Duke, Lord* or *Lady.*

Introduction and History

The second category of fruit growing in Lakeland involved the farms, most of which had a few trees for culinary purposes. The most common and popular variety on the farms in the 19th century was *Keswick Codlin*. The term 'codlin' dates back to around 1600 and is used for an apple picked green which quickly and easily baked to a pulp within a dumpling. This could then be eaten for lunch in the fields, after a slab of bread and cheese. Another widely planted farm variety was *Scotch Bridget* which, as the name implies, originated north of the border, but was popular throughout the north-west because it could be used at harvest as a late cooker, and then stored whereupon its acidity would decline and it could subsequently be eaten as a dessert apple.

Old map records show the locations of orchards in Cumberland, Westmorland and north Lancashire, of which there were a surprising number in the 19th century, especially on the sheltered Kent estuary coastline area, the Lyth and Winster Valleys (including damsons of course) and the Eden Valley. Their fruit was sold locally and in market towns such as Kendal, Appleby and Penrith. Today only a few remnant orchards remain, but they are supplemented by the country houses, whose orchards and gardens are described later in this book.

Introduction to Cumbrian Apple Varieties

THERE ARE AROUND twenty different Cumbrian apple varieties described in this book. Can't we be more precise? No, for two reasons.

Firstly, when is a variety accepted as a variety? Several are well known, and have been for decades if not centuries, such as *Keswick Codlin* and *Duke of Devonshire*. But new varieties are being proposed based on seedlings found or raised. Our definition of a variety therefore is one which has been widely grown, and considered to be worth growing, and for which scion wood is widely available for propagation. This means that some proposed new seedling varieties are considered to be 'in the pipeline' and under evaluation by groups such as Acorn Bank, South Lakeland Orchard Group (SLOG) and individual fruit growers; thus not enough is yet known about them to merit a full variety description, so a dozen such are listed separately.

Secondly, there is some confusion as to whether some varieties may in fact be duplicates. For example, *Royal* may be identical to *Churn Lid*, both from the Lyth Valley, and furthermore, both may be identical to the Scottish variety *Cambusnethan Pippin*, which, Bunyard asserted in 1920, "...is much appreciated in Northern England." The obvious and quick way to solve this conundrum would be DNA testing but this technology is not yet widely available. The only recourse is therefore to grow all three side by side on the same rootstocks and compare their fruit in due course, which may take up to five years.

The National Collection at Brogdale do have access to DNA testing technology and have applied it to all varieties in the National Collection. This has thrown up some surprises: they have concluded that two Cumbrian varieties, *Fallbarrow Favourite* and *Wheaten Loaves* are one and the same, whilst another tree claimed to be *Wheaten Loaves* is considered to be identical to a French variety. They do admit the possibility

Allington Pippin at Briery Bank orchard, Arnside.

of sampling error, so again SLOG proposes to grow these varieties side by side as described above to decide for ourselves.

Another twist in the definition of what is a Cumbrian variety is the subject of the 1974 county boundary changes. Historically speaking, most of the varieties described below were raised in Cumberland or Westmorland. Two were raised in an area that used to be Lancashire at the time, but since it is part of Cumbria now, it seems appropriate to claim that all are Cumbrian varieties. Historically, and especially in Georgian and Victorian times, some varieties originating elsewhere were found to be suited to the rather damp Cumbrian climate and became popular and widely grown here. They therefore justify inclusion below in a separate group following the Cumbrian varieties.

Meanwhile, the actual geographical origin of many varieties is debatable. Some origins are assumed based on first records from historical sources. For example the apple literature asserts that *Scotch Bridget* was first described in 1851, and that it originated in Scotland. However, recently discovered records suggest that it was known in Cumbria in

1781, so could it have been of Cumbrian origin? There is a need for more historical research into this and many other varieties.

Finally, national recognition of an apple variety is conferred by its inclusion in the National Apple Register, which follows successful evaluation by the National Fruit Trials at Brogdale in Kent to demonstrate that the variety is distinctly different from any other. Thirteen Cumbrian varieties are so listed.

Note regarding photographs:
The photographs showing four entire and two half-section apples are of apples grown at the National Fruit Collection in Brogdale, Kent. They are reproduced here with the kind permission of the National Fruit Collection. Almost all other photos are of apples grown in Cumbria. Any differences between them are a result of the very different climates between Kent and Cumbria.

Cumbrian Apple Variety Descriptions

AUTUMN HARVEST ECD (N)

Supposedly of Westmorland origin and first recorded in 1934 but thought to be older. Early autumn dual purpose apple, pick from end August, for use in September when it sweetens to a dessert apple. Trees are vigorous, reliable and disease resistant. Fruit mainly green, medium/large. Flesh firm, tender, coarse, dry with a subacid flavour. There is some confusion about the exact identity of this variety in the National Collection at Brogdale. DNA testing claims that it is identical to *Reverend W Wilks* (raised 1904, Chelsea). Local opinion is that the two varieties are distinctly different.

Key, 1st letter: E = early, M = mid, L = late;
2nd/3rd letter: C = cooker, D = dessert, CD = cooker/dessert
(N) denotes listing in National Apple Register

BRADLEY'S BEAUTY MCD

A very hardy and disease resistant variety found on the mosses in Witherslack by Mr. Bob Bradley in 1975, thought to be a wilding originating in the mid-20th century. A mid season dual purpose apple which sweetens as it matures. Pick from October, for use to year end. Trees are upright and vigorous, possibly triploid. Fruit large, conical with deep and open eye, ripening from light green to yellow with red flecks. Flesh crisp, drops to a sauce on cooking.

CARLISLE CODLIN MC (N)

Originated in Carlisle, first recorded 1816. Mid-season cooker, pick from September for use in Sept/Oct. Trees are moderately vigorous, upright-spreading but susceptible to scab. Fruit yellow, small-medium, turning greasy. Flesh soft white, juicy, sharp and fruity. Cooks to a soft white juicy purée. Bunyard (1920) claimed it to be an apple 'of no value'.

CHURN LID MDC

A flat shaped dual purpose apple found in an old orchard called Borderside near Crook in South Cumbria. It is similar, and may be identical, to an apple called *Royal* from Whitebeck orchard in the Lyth Valley.

DUKE OF DEVONSHIRE LD (N)

This variety was raised in 1835 by Mr. Wilson, gardener to the Duke of Devonshire at Holker Hall. Introduced in about 1875 it was a popular apple in Edwardian times. Late dessert, pick October, will keep up to March, flavour improves in store. Good cropper. Trees are moderately vigorous, spreading and resistant to scab. Fruit dull green with light russet, turning yellow as they ripen. Flesh firm, fine-textured, creamy white, juicy with a rich, sweet-sharp fruit drop nutty flavour. Bunyard called it "quite indispensable for late use."

FALLBARROW FAVOURITE EC (N)

Early season cooker, rediscovered in Mr. Gibson's orchard at Tarnside near Crosthwaite. Name implies a connection with Fallbarrow Park at Bowness. First recorded in 1936 and exhibited from Westmorland in 1946 but is probably much older. It is a flattened, medium sized, yellow-green apple with a warm amber flush. Flavour is rich and mellow with a slightly sweetish tang. It falls almost to a purée when cooked.

FORTY SHILLING MD (N)

Origin Thursby, near Carlisle 18th century. Mid season dessert; pick from September for use through October. Trees are vigorous and good croppers which do well on difficult sites in high, wet and cold regions. Fruit medium sized with attractive red stripes. Flesh creamy white, soft, juicy with a rich sweet-savoury flavour.

GREENUP'S PIPPIN MC (N)

Found in the garden of shoemaker Mr Greenup in Keswick in the early 18th century and introduced in 1769. Widely grown throughout the border counties in the 19th century. Mid-season cooker; pick September for use through to November. Trees are vigorous, hardy and upright-spreading, disease resistant and will tolerate adverse conditions. Fruit medium large, yellow/green with bright red flush. Flesh creamy white, firm, crisp and juicy, cooks to a creamy purée with a sweet-sharp flavour. Synonyms: many, including *Cumberland Favourite, Green Rolland, Red Hawthornden, Yorkshire Beauty*.

JOHN HUGGET ECD (N)

Raised 1940 by John Hugget in Grange-over-Sands from an *Allington Pippin* cross. Pick August/September for use through October when it sweetens to a dessert apple. Trees are moderately vigorous. Fruit medium to large sized, red over yellow. Flesh sweet and rich. Cooks to pale cream purée, sweet and rich with lots of flavour.

KESWICK CODLIN EC (N)

Found growing on a rubbish heap at Gleaston Castle near Ulverston in the late 18th century. This apple was first recorded in 1793 and became one of the most popular early cookers throughout England in the 19th century. It was introduced by John Sander, Keswick nurseryman, thus the name. Early season cooker mostly picked September for use in Sept/Oct. Prolific cropper even in adverse conditions. Trees are moderately vigorous, upright-spreading and fairly resistant to scab. Fruit are long and angular, pale green-yellow, often with a hairline. Flesh soft, rather coarse-textured, somewhat dry and acid, cooks to juicy cream purée.

LANCASHIRE PIPPIN MC (N)

Despite the name, this variety is thought to originate from Underbarrow, Westmorland, some time prior to 1950. Mid-season cooker, pick end September for use in October. Trees moderately vigorous, spreading, possibly triploid. Good disease resistance. Fruit large with attractive bright red stripey flush. Flesh soft, slightly coarse. Slightly rich flavour but rather bland when cooked.

Lemon Square MCD

A lemon yellow dual purpose apple thought to originate from the Penrith area of the Eden Valley. It was listed in an 1820 catalogue of Alexander Forbes, gardener at Levens Hall Nursery. He described it as above medium size, pale yellow and red, and in season from November to December. This disagrees with Barron's description at the 1883 National Apple congress, when it was exhibited from Arnside. He called it small, early season, oblong, angular, yellow and 'worthless', but it may be a case of mistaken identity because *Lemon Square* is medium to large sized. The variety almost disappeared in the late 20th century but was rediscovered by Chris Braithwaite, head gardener at Acorn Bank. The 'square' part of its name is believed to relate to its use as the apple filling in pastry squares, a local speciality, and the fruit itself does have a 'boxy' shape. It has particularly attractive blossom.

LONGSTART MDC (N)

This apple originated in Westmorland, and was first recorded as 'Long Starts' in grafting records from 1781 in the Lyth Valley, then again in 1815 when four were planted on the common at Grange-over-Sands. It was a favourite cottage garden apple around Lancaster and in Westmorland in the 19th century. It is said by Mr D. Holmes of the Lyth Valley to have been grown in every orchard in his district in the past. Its name probably comes from its unusually long stalk. Reliable and healthy mid-season dual purpose apple. Pick from September for use in September to October. Trees are weakly vigorous, spreading. Flesh is soft white, savoury, brisk and juicy with sub-acid flavour.

NELSON'S FAVOURITE MC (N)

A round red striped apple that is thought to originate from Kendal, Westmorland, some time prior to 1958, but probably much older. Mid season cooker, pick from September for use Sept/Oct/Nov. Trees are weakly vigorous. Flesh greenish white, firm, rather coarse. Keeps shape when cooked with good sharp fruity taste. Nelson is thought to have been a local preacher.

RANKTHORN MD (N)

Believed to have originated as a seedling growing in a wood called Rankthorn in Cartmel Fell, Westmorland area in the 19th century. The last known local tree grew at Whitebeck Farm in the Lyth Valley and although this tree has died, a graft was taken and the resultant young tree has been planted there. Mid-season dessert, pick from October for use though November. Trees vigorous. Fruit beautifully coloured with bold red stripes. Flesh soft white, sharp and juicy.

ROYAL MDC

A flat striped dual purpose apple from Whitebeck Farm in the Lyth Valley. Its appearance, purpose and season of use all suggest it is identical to *Churn Lid*.

TAYLOR'S FAVOURITE MC

Taylor's Favourite is a mid-season cooking apple which grows in Mr. Holmes' orchard at Whitebeck Farm in the Lyth Valley. It was, as its name says, the favourite apple of the Taylor family, who took the fruit to sell at local markets in days gone by. There are three trees which were originally planted in 1879 which are disease free and still produce a good crop of healthy colourful autumn cooking apples. It is considered the best autumn cooking apple for the area and has stood the test of time. When cooked it softens but keeps its shape completely, developing a rich flavour.

CUMBRIAN APPLE VARIETY DESCRIPTIONS

TIFFEN MC (N)

This apple is thought to originate from the north-west, and probably Westmorland, sometime prior to 1883. Mid-season cooking apple, pick from September and use though October. Trees are moderately vigorous. Fruit large, green turning yellow with red flush. Flesh white, soft and coarse with a sub-acid flavour. Cooks to sharp juicy purée.

WHEATEN LOAVES EC

An early cooker handed in to an Apple Day by Mr. D. Holmes from an old tree on his farm at Whitebeck in the Lyth Valley. Whitebeck has been in his family since 1747 and has several orchards. Fruit was taken to market up to the 1960s. The tree has dark green leaves and is scab-resistant. There is some confusion about the exact identity of this variety in the National Collection at Brogdale. DNA testing claims that two trees are identical to *Fallbarrow Favourite* while a third tree is claimed to be identical to a French variety, *Transparente de Croncels*. SLOG plans to grow all three varieties close together to investigate further.

Varieties widely grown in Cumbria

MARGIL LD (N)

This variety is thought to have originated in France and first imported to England in the late 17th century. It was grown from the 18th century in the Crosthwaite area. 'Margells' were recorded as being planted at Grange-over-Sands in 1812. A popular late dessert in Victorian time, pick October for use through December. Trees are weakly vigorous, heavy cropper and hardy but prone to canker. Fruit is small to medium size with orange/red flush and stripes over gold with some russet. Flesh is cream, firm, rather dry with a rich aromatic flavour but does require sunlight for full flavour development.

SCOTCH BRIDGET LC (N)

This variety it thought to originate in Scotland, first described by Hogg in 1851, but records show that a 'Britchet' was grafted in 1781 in the Lyth Valley, and 'Britcharts' were planted in 1812 at Grange-over-Sands. It was widely grown in commercial orchards in the 19th century throughout the north west because of its reliability in the local climate, and was the most popular late cooking apple in the area, before *Bramley's Seedling* took over. Late season cooking apple. Pick from October for use through December, when it sweetens to a dessert apple, especially in south Cumbria and in Lancashire. Can be stored up to March. Trees are moderately vigorous, triploid, upright-spreading and very hardy producing heavy crops even in adverse conditions. Fruit are medium to large sized, ribbed and crowned, with orange/red colour spreading over a green/yellow base. Flesh crisp creamy-white, juicy and rich. Does not break up on cooking and has delicate flavour. Synonym *White Calville*. A variant known as *Lancashire Scotch Bridget*, claimed to be slightly but distinctly different, was widely grown around the Preston area, dating from the late 19th century.

Lost Varieties

Various historical records describe old apple varieties which are no longer authentically known today. But from time to time, a 'lost variety' turns up. The following varieties are listed as having existed at some time in the past in Cumberland and/or Westmorland in the hope that someone may know of a tree of the same or closely similar name.

Varieties exhibited at National Apple Congress, London in 1883: *Dacre* (may be the *Daker Apple* planted at Grange in 1812 and/or the *Daker's Delight* planted at Grange in 1815), *Green Soldier* (grown at Heysham in 1754, Grange 1815 and listed in Levens Hall catalogue 1820), *Jamie Brown, Old Neddie, Prussian Pippin* (planted at Arnside 1776 and Grange 1815), *Queen's Crab, Queenstown.*

Varieties listed in *Apples and Pears: Varieties and Cultivation in 1934*, Royal Horticultural Society, London 1935: *China Square, Cumberland Gilliflower, Harvest Lemon* (from Eden Valley), *Housewife, Lochinvar, Mary Barnes, North Britain Giant, Old Pippin, Oversea Peggy, Poolkeeper, Red Custard, Rev. J. B. Hall.*

Others: *Lowther Castle* – records from Cumberland and Westmorland Farm School at Newton Rigg show it was planted in their orchard in 1902 and yielded a good crop of dual purpose apples in 1909. *Carlisle Castle* – recorded in 1920. *Michaelmas Daisy* – recorded in 1912.

Do you have or know of any of these lost apple varieties? We would like to propagate them to ensure the survival of their unique genetic identity. Please let South Lakeland Orchard Group know and we can graft new trees from them. To contact us see:

http://www.slorchards.co.uk/SLOGcontact.htm

Seedling Varieties

THESE ARE RELATIVELY unknown varieties which have originated as seedlings in Cumbria. Only a few trees yet exist and they are still undergoing evaluation to determine their merit or otherwise. Scion wood is not yet widely available.

Burgh Beauty: from the Old Vicarage at Burgh-by-Sands on the Solway Firth. The tree is shown on old maps from the 19th century so could be a good survivor for our area.

Cockermouth Codlin: a hedgerow tree found growing a mile or so outside Cockermouth. Fruit looks similar to *Keswick Codlin*, as it has distinctive ribs, and a similar lemony colour, but flowers and fruits much later and is round/flat in shape. It is a prolific cropper, most years, harvest in early October. Flesh is soft, melting and has been described as like eating cider! Apples do not keep very long in store.

Daphne's Delight: a dual purpose seedling raised in Grayrigg, north-east of Kendal.

High Head: a seedling apple given in at Acorn Bank during an Apple Day (pictured right).

King's Meaburn: a seedling found growing beside the Chapel at the village of the same name.

Lorton Vale: a mid/late season dessert apple raised recently by Low Stanger Nursery from a *Granny Smith* pip. Tends to crop too heavily and benefits from thinning. Reliable cropper, but susceptible to scab. Flesh is slightly pink some years, very juicy and full flavoured. Skin can be a little bitter, which slightly offsets the usually delicous flesh. Attractive blossom.

Mauld's Meaburn: a seedling found growing at the Old Smithy in this attractive village on the west side of the Eden Valley, and an extraordinarily large early eating apple (pictured right).

Millyard Red: an attractively coloured red dessert seedling apple found growing by the watermill at the National Trust property Acorn Bank near Penrith (pictured below).

Tarnside Red: A mid/late season dessert apple found in Mr. Gibson's Tarnside Orchard near Crosthwaite. The original tree, now dead, was quite old. Fruit is mostly bright red, medium size, crisp and juicy.

More seedlings are continually being proposed and entered into evaluation, such as **Alan's Apple**, **Ritson's Red**, **Roland's Reliable**.

Cumbrian Open Orchards

THERE ARE A surprising number of orchards, or gardens with fruit tree collections, open to public access in Cumbria. They are widely distributed around the county as shown on the map on the following page. Fifteen are described in detail and a further handful of smaller sites receive a briefer mention.

The fruit tree varieties are recorded within the location descriptions below, however individual tree labelling is patchy and often absent. This is because many labels weather and deteriorate, get damaged or lost, and sadly may be taken by the public as a reminder of the name of a variety they particularly like. So if you are keen to view a particular variety at a given location, it is best to seek out the gardener for advice.

Opening times and admission prices are subject to annual changes so please check if possible. The details included here are those for the 2013 season – no responsibility is accepted for errors or changes.

The Apples and Orchards of Cumbria

Key:
1 Acorn Bank; 2 Ashmeadow; 3 Brantwood; 4 Cylinders Estate;
5 Dalemain; 6 Grange-over-Sands Community Orchard;
7 Hutton-in-the-Forest; 8 Lakeside Hotel; 9 Leighton Hall;
10 Levens Hall; 11 Mirehouse; 12 Rydal Hall; 13 Sizergh Castle;
14 Winderwath Gardens; 15 Wordsworth House and Garden.

1. ACORN BANK

The name of this National Trust property, originating from the ancient oak woods surrounding the Crowdundle Beck, dates from 1597, and the red sandstone manor house dates from the 16th century, though most of it is 18th century. It passed into National Trust ownership in 1950 along with gardens and almost 200 acres of land.

The watermill has been restored and the woods bloom in spring with over 60 different varieties of daffodils, narcissus and wood anemonies.

Main orchard in walled garden: Bramley in foreground and clock tower

The gardens boast the north of England's largest collection of medicinal and culinary herbs which extends to 250 species and varieties.

The orchards are also a notable feature. The oldest trees within the walled garden date from the 1930s and include a *Blakeney Red* (Pear), *Cambridge Gage*, Quinces, Medlars and Mulberry.

The main orchard of 26 apple trees was planted from 1972 as half standards on MM111 and consists of the following varieties chosen for their suitability for the northern climate: *Oslin, Lord Hindlip, Carlisle Codlin, Charles Ross (2), Kidds Orange Red (2), Lord Lambourne, Norfolk Beefing, Keswick Codlin, French Crab, Bramley's Seedling (2), Egremont Russet (2), James Grieve, Red Ellison, George Cave (2), Discovery, Grenadier, Irish Peach, Scotch Bridget, Bardsey, Lemon Square and Dumelow's Seedling.*

A new orchard of half standards on M25 was planted from the year 2000 containing: *Lemon Square, Hawthornden, Tiffen, Nelson's Favourite, Forty Shilling, Flower of the Town, Ribston Pippin, Newton Wonder, Rankthorn, Ladies Finger of Lancaster, Scotch Bridget, Autumn Harvest, Pitmaston Pineapple, Lancashire Pippin, Greenup's Pippin,*

Golden Noble, Irish Peach, Blenheim Orange, Edward VII, Lady Sudeley, Ellison's Orange, Hargreaves Greensweet, Yorkshire Cockpit, Proctor's Seedling, Nancy Jackson, Duke of Devonshire, Ashmead's Kernel, George Neal, Grenadier, Bramley's Seedling, Bradley's Beauty, Howgate Wonder, Mere de Menage, Maulds Meaburn, Golden Spire, Yorkshire Greening.

More recently, four rows of cordons on MM106 have been planted containing about twenty varieties per row including:

[1] *Warner's King, Monarch, Bismarck, Peasgood's Nonesuch, Bountiful, Alexander, Ben's Red, Kidd's Orange Red, Mother, Queen, Annie Elizabeth, Laxton's Superb, Lord Derby, Rev. W. Wilks, Norfolk Beauty, Edward VII, Golden Delicious*

[2] *Churn Lid, Wyken Pippin, High Head, Keswick Codlin, Wheaten Loaves, Taylor's Favourite, Desmond's Rankthorn, Fallbarrow Favourite, Greenup's Pippin, Court of Wick, Coul Blush, Cox Pomona, Adam's Pearmain, Stirling Castle, Royal Jubilee, Burgh Beauty, Crimson Bramley, Lane's Prince Albert*

[3] *Discovery, Katy, Fiesta, Worcester Pearmain, Orleans Reinette, Gladstone, Jonagold, Tydeman's Early Worcester, Laxton's Fortune,*

The colourful Lady Sudeley at harvest time

Ashmead's Kernel, Suntan, John Standish, King of the Pippins, Gravenstein, Pixie, Red Devil, Scrumptious, Meridian, Limelight, Greensleeves, Redsleeves, Red Falstaff

[4] *Bountiful, Cox's Orange Pippin, Arthur W. Barnes, Ellison's Orange, Longstart, Kerry Pippin, Gascoyne's Scarlet, Braeburn, Tom Putt, Lady Henniker, Red Gravenstein, Emneth Early, Court of Wick, Beauty of Moray, Fearn's Pippin, Allington Pippin*

In total this comes to over 100 different apple varieties which must be the biggest collection of apple varieties in any public Cumbrian orchard! In addition, a further two rows of cordons are planned for 2013. As if that were not enough, there are some tall Perry Pears (*Blakeney Red and Thorn*), fan-trained gooseberries (*Invicta and Greenfinch*) and ten varieties of rhubarb in the new orchard, plus quince (*Portugal and Vranje*) and whitecurrant (*White Transparent*) in the herb garden.

The top apple event of the year at Acorn Bank is undoubtedly 'Apple Day' in mid-October which pulls in crowds of well over a thousand to enjoy a host of activities such as longest apple peel competition, apple bobbing, apple bowling, apple shy, children's workshops and treasure hunt, archery, entertainment, Morris Dancers, storytellers, craft demonstrations, rural stalls and exhibitions; not to mention South Lakeland Orchard Group and North Cumbria Orchard Group stands, the Northern Fruit Group apple identification and advice team, plus the Great Cumbrian Cider Challenge!

In addition, Head Gardener Chris Braithwaite, who recently celebrated 30 years at Acorn Bank, runs pruning and grafting courses which are both educational and entertaining.

For more information see: http://www.nationaltrust.org.uk/acorn-bank/ Open daily except Tuesday from mid-March to early November, 10am-5pm. Admission £5 (National Trust members free).

Location six miles east of Penrith, one mile east of A66, one mile north of Temple Sowerby, follow brown signs.

Postcode for Satnav: CA10 1SP. Telephone: 017683 61893

2. ASHMEADOW

Ashmeadow is a small remnant orchard of about 36 trees within the five acre Ashmeadow Woodlands at Arnside on the south bank of the Kent estuary. It is owned by the Barnes Charitable Trust, which was set up in 1990 to manage the Ashmeadow estate in the interests of the environment, the local community and visitors.

The orchard is mostly apples, with a few pear, plum and walnut trees of ages ranging from newly planted up to over 50-years-old. Phil Rainford's research shows that orchards have been grown at Ashmeadow for over two centuries. It was described as, "a remarkable orchard for bearing well" in the 1770 Beetham Repository, while the *Cumberland & Westmorland Gazeteer* of 1829 said, "the plantations about Ashmeadow are in a thriving condition and the fruit trees extremely luxuriant" (see *Orchards of the Arnside and Silverdale AONB*).

The current variety and rootstock identities have been mislaid but the trustees hope to retrieve them from their records. Six young apple trees planted winter 2012 include *Kane's Seedling*, a Nottinghamshire variety previously popular locally which had been lost but rediscovered by Phil Rainford, plus some unidentified local varieties all grafted by Phil. The

trees are unsprayed and receive no fertiliser but are in reasonable condition except for some canker and scab. The orchard floor is grass with wild daffodils and lesser celandine. The grass is mown and the trees enjoy a sunny yet sheltered situation. Some of the older trees have been overgrown by brambles and ivy in the past but volunteers from the Arnside and Silverdale AONB have progressively cleared the site.

Access is via a public footpath from Silverdale Road which runs through the orchard and continues through the entire woodland area. Beyond the orchard is a substantial walled garden. The far end contains some allotments but most of it is managed as a wild flower meadow, containing orchids, poppies, cranesbill, knapweed, ox-eye daisy and self-heal.

Lower down the hill is another old orchard of about eight trees. Visitors are encouraged to pick the fruit in season. If you happen to be in or near Arnside and have some time to spare, Ashmeadow Woodlands and orchard are both worth a visit as a round tour from the Promenade following the paths shown on the plan page of the website. There are several information boards to guide you round the woodlands, and to explain each area. The website is: www.ashmeadow-woodlands.org

Directions: Leaving Arnside Promenade, go 200 yards up Silverdale Road. Just before Redhills Road a pedestrian gate on the right with public footpath sign gives access to the orchard (just above Ash Meadow Lodge and opposite High Bank). The woodlands can also be entered by public footpaths from the western end of the Promenade which lead to the orchard. Entry is free at all times.

3. BRANTWOOD

Brantwood was the home of the Victorian writer, artist, critic, social reformer and conservationist John Ruskin from 1872 until his death aged 80 in 1900. The house contains memorabilia, art and exhibitions and hosts an events programme including classical concerts. The estate comprises 250 acres of gardens, pastures and woodland stretching from lakeside to open fell. There are eight different gardens, which continue Ruskin's own experiments in horticulture and land management.

Close to the zig-zag path are two apple trees which date from Ruskin's time; a *Galloway Pippin* and a *Bramley's Seedling*, both of which, despite being around 120-years-old, have been rejuvenated by hard pruning. Closer to the house, in the Professor's Garden, is a row of cordons planted on M26 in 1992 using Victorian varieties that are typical of what would have been grown in Ruskin's day. Looking uphill, the seven dessert varieties on the left are *White Transparent, American Mother, Brownlees Russet, Ribston Pippin, Pitmaston Pineapple, Cornish Aromatic* and *Roundway Magnum Bonum*.

On the right are seven culinary varieties, *Norfolk Beefing, Lord Derby, Smart's Prince Arthur, Lane's Prince Albert, French Crab, Golden Noble* and *Belle de Boskoop*. Lower down by the 'Jumping Jenny' restaurant are four young stepovers: *Bradley's Beauty, Dumelow's Seedling, Proctor's Seedling* and *Greenup's Pippin*. The orchard, planted in 1989, is lower down on the lake side of the road. Although it comprises only nine trees, it is full of interest.

It is here that Head Gardener Sally Beamish is developing an innovative orchard management system based on biodynamic principles. These ideas originated from Rudolf Steiner in 1924, who integrated the principles of organic farming with 'natural preparations' and the astrocalendar to advocate a holistic approach to agriculture. A descending lunar rhythm is believed to make sap flow less active, accordingly winter pruning is carried out during such periods.

Entering the orchard, the first of the twenty or so year old trees is a vigorous half-standard *Keswick Codlin* on MM106, then a *Ribston Pippin*, two *Beauty of Bath, Grenadier*, two *Egremont Russet, Edward VII* and *Court Pendu Plat*, mostly bush trees, some on M26, some on MM106. The trees originated from nurseries as diverse as Deacons on the Isle of Wight and Tweedies of Dumfries, some arriving by boat! The *Court Pendu Plat* had suffered from canker, but under the biodynamic regime it has recovered; the scars are still visible but no longer sporulating.

However, early signs of canker have now appeared on the *Keswick Codlin* and *Beauty of Bath*, so although a battle has been won, the war is not yet over. The orchard is grassed down and mown; despite this the trees are vigorous and set too much fruit which then needs to be hand thinned: on the *Keswick Codlin* in particular this amounts to a barrowful of surplus fruitlets. The fruit, when harvested, is all made use of; the best quality sold fresh, the next grade used for cooking in the Jumping Jenny Café, and the surplus juiced and sold in the bottle so nothing is wasted.

The zig-zaggy path has two espaliered pears, a heavy-cropping *Black Worcester* and a rather shy *Jargonelle*, whilst against the south-facing wall of the house is a fig, fruiting happily within a confined rootspace. Sally has been at Brantwood over twenty years, during which time she has led considerable restoration of the garden and continues to initiate new ideas not least the biodynamic project in the lakeside meadows and adjoining orchard. The views over Coniston Water and beyond to the Coniston Old Man range are magnificent, yet it is off the beaten track, and surprisingly tranquil compared to most Lake District beauty spots.

So, if you fancy a spot of volunteer work in a peaceful location with unrivalled views, and/or if you're interested to learn more about biodynamic apple growing – this, probably the only biodynamic orchard in Cumbria, is the place. Website: www.brantwood.org.uk Winter openings (mid-Nov to mid-March) are Wednesday-Sunday 10.30am to 4pm. Summer open daily mid-March to early-Nov 10.30am to 5pm, gardens admission £5.50. Location 2½ miles from Coniston on the east side of Coniston Water. The best way to travel is by boat across the lake from Coniston, either on the National Trust's steam yacht *Gondola* or Coniston Cruise's solar-electric launch, both of which arrive at Brantwood's own jetty at the foot of the orchard. Postcode for Satnav: LA21 8AD. Telephone: 015394 41396

4. Cylinders Estate

The survival of a 70-year-old orchard in Elterwater is due to the Dadaist German artist Kurt Schwitters and to Ian Hunter and Celia Larner, whose arts charity the Littoral Trust purchased Cylinders Estate including the famous Merz Barn in 2006. At the time that Kurt Schwitters used the barn as a studio, the estate was owned by Harry Pierce, a local landscape gardener. Pierce planted 43 apple trees in January 1943, of which about two dozen still survive as mature semi-standard trees. Varieties include *Bramley, Worcester Pearmain, Lord Derby, Annie Elizabeth, Allington Pippin, Beauty of Bath*; and the crabs *John Downie* and *Cheale's Crimson*.

The trees grow on a steep slope of up to 45° covered in a herb-rich meadow which is scythed in July. Many of the trees are in remarkable health considering their damp and shaded situation with high annual rainfall resulting in moss-covered branches with varied lichen. Not surprisingly there is scab and some canker but no sign of insect pests apart from a little capsid damage. After 70 years, a healthy equilibrium must have built up between pests and predators.

In addition there is a row of myrobalan plum, and some damsons. The estate stretches for five acres and includes some interesting big-leaf rhododendrons and specimen trees. Schwitters fled Germany in 1937 after his modernist art was branded as 'degenerate'. The Merz Barn building still stands much as Schwitters left it on his death in 1948.

The Littoral Trust is working to restore the Merz Barn and the Cylinders estate so that the site can be made accessible for the general public, schools, artists and scholars. However this is dependant on funding which at the time of writing is uncertain. At present the site is undergoing restoration and upgrading and so is not open to the public on a permanent basis.

It does however open at Easter weekend, August Bank holiday, and for the annual Kurt Schwitters Autumn Schools (held during the first weekend in October). Other opening times are announced on the website. Location: On B5343, ½ mile west of Elterwater, entrance to Cylinders is opposite Langdale Hotel, on the road to Chapel Stile. Postcode for Satnav: LA22 9JB. Website: www.merzbarn.net. Tel: 015394 33046.

5. DALEMAIN

Dalemain, despite its Georgian façade, contains Tudor and mediaeval elements. Already famous for its hosting of the World's Original Marmalade Awards and Festival which attracts over 1,000 participants, its most recent claim to fame is winning the 2013 Garden of the Year award sponsored by the Historic Houses Association. Originally the site of a fortified tower during the reign of Henry II, it has been home to the Hasell-McCosh family since 1679. Situated just north of Ullswater, it is surrounded by extensive gardens which contain up to 30 different heritage apple varieties.

About twenty trees line the Apple Walk in the main garden, one row of standards beside the path and another row as espaliers against a south facing wall. Three others are below the terrace, and another four are situated randomly in the shady Low Garden. Further trees are in the Kitchen Garden. Tree age is unknown; Mrs. Hasell-McCosh thinks some may date from the 19th century, while some may be early 20th century plantings based on the varieties used.

CUMBRIAN OPEN ORCHARDS

Above, the Dalemain Apple Walk and below trees in the walled garden.

43

Most trees are still healthy and vigorous, carrying good crops despite considerable competition from shrubs and the extensive collection of roses in the borders. Varieties include: *Ribston Pippin, Allington Pippin, Lord Lambourne, Grenadier, Lane's Prince Albert, Laxton Superb, Worcester Pearmain, Newton Wonder, Beauty of Bath, Peasgood's Nonesuch, Charles Ross, Keswick Codlin* and an as yet unidentified old tree which may be a *Downton Pippin*. The fruit is used for cakes, tarts etc. in the tea room. A particularly interesting feature is the fruit store, a purpose built square building at the top end of the garden which dates from the 16th century (pictured above).

Gardens open Sun-Thurs 10.30am-5pm late March to end Sept, (4pm to end Oct) then 11am-3pm in winter (closed mid-Dec to mid-Feb). Postcode for Satnav: CA11 0HB. Tel: 017684 86450. Admission £7. Website: www.dalemain.com Directions: 3 miles south west of Penrith on A592.

6. GRANGE COMMUNITY ORCHARD

Grange has been an apple growing area for many years as evidenced by the Ordnance Survey map of 1850. One such orchard was in the grounds of Yewbarrow Lodge, but was cleared in the Victorian era to create the shops and cafes of Yewbarrow Terrace, just south of the B5277/B5271 roundabout opposite the Ornamental Gardens.

In 1997/98, a new two acre orchard was planted on an adjacent piece of land, originally a paddock, which had been left to the people of Grange by Colonel Porritt, the last owner of Yewbarrow Lodge. The objective of Grange Community Orchard was to create a collection of different varieties of a range of fruit trees including local varieties. It

Above, Grange Community Orchard and below Sunset apples.

now contains over 30 different apple varieties plus pears, plums, damsons, cherries, medlars, quince, hazel and mulberry. Tree labels indicate the variety, origin, date of introduction, etc. Several trees were sponsored by local people.

Local apple varieties are *John Hugget* (Grange's own variety, a sweet, rich dual purpose apple raised in 1940), *Keswick Codlin, Nelson's Favourite, Ladies Finger of Lancaster* and *Duke of Devonshire*. Others are *Annie Elizabeth, Ashmead's Kernel, Beeley Pippin, Belle de Boskoop, Blenheim Orange, Bramley's Seedling, Charles Ross, Crispin, Egremont Russet, Ellison's Orange, George Cave, Golden Noble, Howgate Wonder, James Grieve, Lord Derby, Pixie, Ribston Pippin, Rival, Scotch Bridget, Scotch Dumpling, Sunset, Suntan, Warner's King, White Melrose* and *Wild Crab Apple*.

Pears: *Catillac* and *Hessle*. Plums: *Denniston's Superb* and *Myrobalan, Greengage*. Damsons: *Merryweather, Lyth Valley, Prune, Shropshire Prune* and *Westmorland Damson*. Medlars: *Dutch* and *Nottingham*. Quince: *Meeches Prolific* and *Pear-shaped quince. Black Mulberry.*

Grange Civic Society has managed the orchard since 2003. The grass is cut by South Lakeland District Council, and Grange Civic Society organises pruning and maintenance of rabbit-proof wire mesh guards around the trees. The orchard is managed organically and wildlife is encouraged by allowing a different part of the orchard to grow as a wildflower meadow each year, by planting buddleia to encourage butterflies and also by planting native hedging such as hawthorn, hazel and sea buckthorn along the lower wall against Main Street.

Recent initiatives have been interpretation boards placed at both entrances from Main Street, and information leaflets which are available from the nearby Tourist Information Centre in the Victoria Hall. The recent planting, information leaflets and boards, and wire mesh guards were all financed by a grant from Trans Pennine Express and the Forestry Commission.

The orchard is permanently open to visitors and has a public footpath through it with links to woodland walks in Yewbarrow Wood and Hampsfell. Three SLOG members, Shirley Leaver, Ron and Judith Shapland do most of the work in the orchard on behalf of Grange Civic Society. One of their biggest problems was people picking the fruit too early, so the tree labels now indicate the appropriate picking time for each variety. Entry is free at all times. Directions: Near Grange Railway Station, just south of the B5277/B5271 roundabout, opposite the Ornamental Gardens.

7. HUTTON-IN-THE-FOREST

Hutton-in-the-Forest claims to be the legendary Green Knight's Castle in the Arthurian tale of Sir Gawain and the Green Knight. Its forest was the second largest Royal Forest in mediaeval England. The oldest part of the current house is the Pele Tower, dating from the 14th century, built as defence against the Scots.

The highlight of the gardens and grounds is the 18th century walled garden. The walls were built in the 1730s and were then planted with peach, apricot, plum, pear and apple trees. Yews were added, both here and on the terraces around the house, in the 19th century to reflect the revived interest in topiary by the Arts and Crafts movement. The walled garden has been successively an ornamental Dutch Garden, a kitchen and cutting garden and, during the Second World War, a market garden.

The current layout has evolved over the last 60 or so years although many of the fruit trees are much older. There are 75-year-old apple, pear and plum trees espaliered or fantrained against the south and west walls, and three Morello cherries in an adjacent courtyard. The senior resident tree is a 120-year-old *Bramley* grown as a free-standing espalier which still carries a huge crop of fruit. Other espaliers continue an east-west line from the *Bramley*. Going south from the *Bramley* toward the house are a line of free-standing apple trees. In total there are about 25 apple trees, nine pears, four plums and gages, the three cherries and a damson.

Rootstocks are unknown but clearly vigorous, so possibly M2. Gardener Kevin Caddy says they receive no fertiliser except for three inches of mulch over all the borders from which they will draw nutrients. There is some scab, but not enough to require spraying, also some canker but the trees' vigour enables them to outgrow it. The major pest is woolly aphid, typically on the shady undersides of main limbs which can be rubbed out with a brush from time to time.

Pruning is mostly done in the winter simply for reasons of spreading the workload against priorities throughout the year, but the 120-year-old *Bramley* has to be summer pruned also, otherwise it would be impossible to manage because of its vigour. The other free-standing trees are also summer pruned because they can be done quickly from ground level. The espaliers against the walls involve ladder work which takes much longer. Kevin is a genuine fruit man, having started working on fruit in Cornwall before taking a job in propagation with the famous but now sadly demised nursery business, Scott's of Somerset. At Scott's he would t-bud thousands of roses in July before moving on to chip bud apples, pears and stone fruit in August; in all cases working on rootstocks in the ground which meant working doubled over.

His record was chip budding over 2,000 apples in one day. This was done by using a quick raffia tie to hold the bud in place which was subsequently secured with tape by a student following behind. He reckoned on about a 92% success rate, so there were still a few hundred failures to be grafted over in the new year in addition to a few thousand more grafts; in total this came to about 40,000 buds and grafts per season.

A fair proportion of the apples at Hutton-in-the-Forest are earlies which have to be picked and used as they ripen. The fruit is used for a range of purposes, some in the kitchen and tea room, some (e.g. *Lady Sudeley*) for decorative purposes in the public rooms of the hall, some for juicing, while *Egremont Russet* is Lady Inglewood's favourite. There is an apple store upstairs in an outbuilding where *Cox* and *Bramley* are stored in single layers on paper over wooden trays to Christmas provided the mice don't get them first.

Amongst the varieties, the *Cox* seems to do fairly well without too much scab or canker despite the Cumbrian climate, although the rainfall at Hutton-in-the-Forest is lower than in most of Cumbria. The biggest disappointment is the *Crawley Beauty* which rarely crops, in fact rarely has blossom. One of the gages has silver leaf which has required the removal of its central leader, it now seems to be spreading to other branches so will probably be removed in the next few years. Otherwise the trees are in good health for their age.

Varieties – APPLES: *James Grieve* (2), *Worcester Pearmain, Laxton's Fortune, Laxton's Superb* (2), *Cox* (2), *Duchess of Oldenburg, Lady Sudeley, Bramley* (2), *Charles Eyre, Charles Ross, Crawley Beauty, Egremont Russet, Ellison's Orange* (2), *Discovery, Sunset, Rosemary Russet, Lord Derby, Beauty of Bath, Rival.*

PEARS: *Conference* (3), *William's Bon Chretien, Concord, Winter Nelis* (3), *Beurre de Comice.*

STONE FRUIT: *Victoria plum, Oullins Golden Gage*, other gage, *River's Early Prolific, Merryweather Damson, Morello Cherry* (3).

Website: www.hutton-in-the-forest.co.uk. Gardens open end March to end October daily except Saturdays 11am to 5pm. Gardens and grounds admission £6. Location: three miles north west of M6 junction 41 on B5305. Postcode for satnav: CA11 9TH. Tel: 017684 84449

8. LAKESIDE HOTEL

How many fruit trees does it take to count as an orchard? The Lakeside Hotel has only eleven, but the main reason for including this site here is that this is possibly the best example of publicly accessible espaliered apples and pears in Cumbria. The hotel is situated at the southern end of Lake Windermere, surrounded by gardens running down to the lakeshore.

The hotel car park is immediately north of the hotel. As you walk out of the car park, a line of eight espaliered apples is situated diagonally to your left. Each variety is labeled with an Alitag label (www.alitags.com) showing planting date. The varieties are: *Scotch Bridget, Duke of Devonshire, Old Pearmain, Golden Spire, Egremont Russet, Keswick Codlin, Ladies Finger of Lancaster* and *Carlisle Codlin*.

The trees were planted by SLOG founder member Dick Palmer on MM106 rootstocks. They stand above the hotel's pool room, in only 15cm depth of topsoil. Most are now 10-years-old and have been summer pruned annually with four pairs of branches carrying plenty of fruit spurs. Gardener Brian Gardner-Smith explained that they receive an annual spring dressing of mushroom compost and also benefit from the lawn fertiliser, but nothing else since pests and diseases (except a little scab) are minimal. The open sunny aspect and good airflow are probably beneficial factors.

Round on the west side or roadside of the hotel, there are three espaliered pears (*Louise Bonne de Jersey*), an early flowering dessert variety planted in 2003. These are managed in the same way on four wires and look healthy and vigorous. Brian says they crop well but suffer more scab than the apples.

The Lakeside Hotel opens its gardens for charity under the National Gardens Scheme usually on one Wednesday in mid-June and one in late

August, 11am to 4pm, admission £5. For anybody who wants to see what espaliered apples and pears should look like, the Lakeside Hotel is well worth a visit.

Website: http://www.LakesideHotel.co.uk. Location: One mile north of Newby Bridge, follow brown signs for Windermere Lake Cruises. Postcode for Satnav: LA12 8AT. Tel: 015395 30001.

9. LEIGHTON HALL

Although Leighton Hall is not in Cumbria, it is only just over the border into Lancashire. It is on the edge of the Arnside and Silverdale Area of Outstanding Natural Beauty (AONB) and nestles into beautiful parkland backed by views of the Lake District Fells. The hall has an orchard which was marked on 1846 maps. However all that is left are two 70-year-old apple trees, one identified by Phil Rainford as *Warner's King*, and the other unknown, possibly a seedling since there is no sign of a graft.

There are also three younger trees, about 20-years-old. More interesting is the Farm Orchard which dates back over 150 years. This contains mainly cherry trees. Three are old but edible, of unknown variety, plus some ornamental cherries and a couple of apple trees. The major interest is the 19th century walled garden which has the remains of an orangery on the south wall.

This used to have espaliered fruit trees around the walls in the 19th century but they had all gone before the Second World War. There is now a vigorous free-standing 65-year-old *Crimson Bramley* (pictured on page 53), and an old

Lancashire Pippin.

unknown but juicy pear against the wall.

Gardener Steven Lewis decided to start replanting apples against the wall a few years ago. Hilary Wilson and Phil Rainford provided some Lancashire heritage varieties grafted onto MM106. A dozen were planted as cordons on the south facing wall, and more are planned to complete the space available. The trees do not suffer

Duke of Devonshire and mason bee box.

from scab, perhaps due to the lower rainfall here. They are kept weed-free in the border and get a low level of fertiliser. The walled garden also contains a young free-standing *Lancashire Pippin*, four crab apples, a young *Conference* pear and a couple of ornamental cherries whose days are numbered due to honey fungus.

Deer, squirrels and rabbits are all problems in the old orchard, but are excluded from the walled garden by keeping gates tightly closed. The twelve cordon varieties are: *Proctor's Seedling, Duke of Devonshire, Keswick Codlin, Ladies Finger of Lancaster, Pott's Seedling, Harvest Festival, Lady's Delight, Hargreaves Greensweet, Hutton Square, Lange's Perfection, John Hugget, Royal George.*

Website: www.leightonhall.co.uk. Open May-September, Tuesday-Friday, 2-5pm. Gardens admission: £4.50. Tel: 01524 734474. Location: halfway between Milnthorpe and Carnforth. Signposted from junction 35A of M6, one mile west of Yealand Conyers. Postcode for Satnav: LA5 9ST

10. LEVENS HALL

Levens Hall is best known for its unique topiary garden, claiming to be the oldest in the world, dating from 1694. Less well known, but nevertheless of significant interest is the orchard containing around 100 trees, mainly apples, but also including pears, crabs, medlars and quince. Head gardener Chris Crowder describes it as a 'Romantic Orchard' because of its eclectic nature, the trees being underplanted with red tulips and crisscrossed with mown grass pathways. Some trees are over 100-years-old, dating from the original planting but the majority have been replanted at varying times since, resulting in a wide range of tree ages and size.

There is a wide range of traditional apple varieties, but the identity of some has been lost. All trees are on unknown but vigorous rootstocks grown as standards or half-standards on approximately twelve feet square spacing. The orchard is unsprayed and unpruned but despite that, most trees are in reasonable health. There is an interesting range of lichens on most trees, whilst some have climbers such as honeysuckle, clematis and rambling rose scrambling over them, though those so encumbered are suffering as a consequence. The apples are harvested for cider which is sold in the Potting Shed Shop, so the orchard pays its way.

A recent initiative has been the planting of medlars and quinces to replace some old diseased trees. Chris considers the medlars have been a great success, their blossom being a particular attraction (as is the quince blossom). Unfortunately the quinces have suffered badly from quince leaf blight which causes the leaves to turn brown and fall in mid-season. This is a disease which is exacerbated by wet seasons which are 'normal' in Cumbria. Annual rainfall at Levens is about 45 inches which must be one of the lowest in Cumbria, but since the gardens are practically at sea level, the relative humidity may be a factor. The worst affected quinces have now been grubbed out leaving only one *Vranje* which, being on the south side of the orchard, may get enough sunshine and airflow to reduce the disease pressure and survive.

Medlar

Another relatively new feature in the garden is a 'Nuttery', which is now about fifteen-years-old. About 40 bushes are grown on two sides of a quadrangle layout shared with the kitchen garden. They provide catkins early before other species show any activity, followed by attractive autumn foliage. The crop is mainly taken by squirrels but the prunings make useful supports in the vegetable garden.

Other fruits grown include vines trained against a wall and figs against the hall, whilst records show that a peach was grown outdoors against a wall during an earlier warmer period. There is also a particularly wrinkly 70-year-old black mulberry which is quite a specimen and looks much older yet carries a good crop of tasty fruit. Chris reckons May is the best time to visit the orchard to enjoy the apple and medlar blossom, complemented by the red tulips beneath.

The orchard includes the following varieties: *George Cave, Bishop Wolsey, Lord Lambourn, Laxton's Superb, Nelson's Favourite, Early Victoria (syn. Emneth Early), King of the Pippins, James Grieve, Norfolk Royal, Katy, Kings Acre Bountiful, Egremont Russet, Ontario Reinette, Bramley, Newton Wonder, Ribston Pippin, Keswick Codlin, Reinette Rouge Etoilee, Worcester Pearmain, Norfolk Beauty, Delicious, Margil, Queensbury, Lady Sudeley.* In addition, SLOG has supplied the following varieties to fill the gaps left by the quinces: *John Hugget, Howgate Wonder, Gravenstein, Warners King, Autumn Harvest, Tom Putt.*

Website: www.levenshall.co.uk. Open early April to early October, 10am-5pm, Sunday to Thursday. Gardens admission £8.50. Location: six miles south of Kendal on A6 just south of A590. Postcode for Satnav: LA8 0PD.

11. MIREHOUSE

Situated on the south east shore of Bassenthwaite nestling beneath Skiddaw, very close to the Lake District Osprey Project observation points in Dodd Wood, with which it shares the car park (charge refundable against entry), Mirehouse is a country manor house built in 1666. Records and maps show the existence of orchards from the 18th century onwards. A Scots Pine planted in 1784 still stands today. The varied collection of rhododendrons are particularly attractive in late spring.

The walled garden dating from about 1780 contains three different orchard plantings. The earliest planting is believed to be over a hundred-years-old, but only three trees remain. One, a *Lord Suffield* is still in good condition, but the other two (a *Newton Wonder* and an unknown variety) may not last much longer due to canker. The second planting is believed to date from the 1950s and consists mainly of dessert varieties fan trained around the south facing wall. Most of these trees are in reasonable health and provide a good crop of fruit.

Variety records have been lost but some have been identified as *James Grieve, Worcester Pearmain, Cox's Orange Pippin* (right), *Lord*

Lambourne and *Egremont Russet*. The third planting was carried out by the present owner in 1996 and consists of a range of heritage varieties grown as free standing standard trees growing in the centre of the walled garden. These are understood to be on M25 rootstock, yet are surprisingly small. However they are mostly in good health, cropping reasonably with vigorous vegetative growth. Exceptions are a few trees in a low damp spot which died. Despite being unsprayed, there is not much scab, the wide spacing ensuring good airflow to avoid excessive humidity. Woolly aphid is a problem on some varieties. List of varieties of fifteen-year-old trees, APPLES: *James Grieve* (2), *American Mother, Northern Greening, Gloria Mundi, Egremont Russet* (2), *Catshead, Broad Eyed Pippin, Bushy Grove, Beauty of Bath, Golden Spire* (2). PEAR: *Doyenne du Comice*. PLUM: *Victoria*.

In the north west corner of the walled garden there are bee hives, one of which houses a colony from Germany which is particularly active. Further north from the house is a remnant plum orchard which includes a greengage still cropping well despite being now overgrown by woodland and *Rhododendron ponticum*. Close by is a huge 200-year-old sycamore.

Egremont Russet.

The grounds extending down to the shore of Bassenthwaite Lake make it an attractive place to visit. Website: www.mirehouse.com. Gardens open April to end October daily 10am to 5pm. Gardens and grounds admission £4 (children £2). Location: three miles north of Keswick on A591. Postcode for satnav: CA12 4QE. Tel: 017687 72287

12. RYDAL HALL

Rydal Hall belongs to Carlisle Diocese of the Anglican Church. The hall, a beautiful old building set in parkland and woodland with attractive countryside views, is used for conferences and courses. The gardens, originally designed by Thomas Mawson, are a unique example of Edwardian gardening in the heart of the Lake District.

After decades of post-war decline, the gardens were restored to their former glory in a major restoration project started in autumn 2005.

Records show that an old walled kitchen garden has supplied fresh produce to the house from the 19th century if not earlier. Part of this area has now been restored as a community vegetable garden, within which an orchard was established at the end of 2006.

There are eighteen mainly northern heritage apple varieties in the orchard, grafted onto MM106 rootstocks and grown as half-standards spaced ten feet apart in three rows. They are vigorous trees grown in an organic regime with only a little scab, but inevitably some canker. Head gardener Ian Turnbull's greatest concern is their tendency to premature defoliation. The soil is fairly well drained loam on a slope, and the trees have been fed with pelleted chicken manure. In addition there are rare breed Maran hens in the orchard which reduce pest and weed competition whilst returning more nutrient.

Rabbits and deer in the surrounding wood are excluded by means of a deer fence. Unfortunately it is totally ineffective against the invading Japanese Knotweed. There are two old apple trees, variety unknown, presumably remnants of the original orchard, one inside and one outside the new orchard and a walnut tree.

The adjoining vegetable garden contains more fruit. A *Keswick Codlin* has been fan-trained against a garden shed at the entrance, an *Ecclestone Pippin* stands just outside the orchard, a *Bardsey* further up, while an *Ellison's Orange* and a *Bramley* have been fan-trained against the back wall. There are also a pear tree, a damson and two apricots, plus hazelnuts, blackcurrants, redcurrants, gooseberries and strawberries.

The glasshouse against the back wall contains a vine, fig and *Royal George* nectarine, all fruiting. Much of the maintenance work such as pruning etc., is done by volunteers and Ian welcomes any new volun-

Keswick Codlin (above) and Ellison's Orange (below)

teers as there is plenty of work to do, in a very attractive location.
 The Rydal Hall orchard is an interesting opportunity to see how a diverse range of old apple varieties prosper in a challenging environment. The eighteen varieties in the orchard are: *Lord Derby, Golden Noble, Edward VII, Charles Ross, Millicent Barnes, Egremont Russet, Kings Acre Pippin, Grenadier, Annie Elizabeth, Darcy Spice, Peasgood's Nonesuch, Newton Wonder, Blenheim Orange, Scotch Bridget, Burr Knot, Pitmaston Pineapple, Withington Welter, Yellow Ingestrie.*
 Rydal Hall Gardens are open free of charge, 365 days a year, from 10am to 4.30pm. Website: www.rydalhall.org. Location: one mile north of Ambleside on A591, take right hand turning marked 'Rydal Hall' on blue sign. The hall is first on the right, through wrought iron gates. Park your car within the grounds, walk toward front door, then take steps up on left, go across road, through gate and up path for 100 yards through wood. The path opens into a field, now used as a campsite and you will see Rydal Community Vegetable Garden through the gate on right. Go through gate and follow short path on right down to the orchard through another gate. Alternatively it is possible to park closer to the orchard by continuing by car up the road past the main gates for another 200 yards until you see a gate on right marked 'Rydal Hall – campsite users only'. Park inside gate and walk across field to the garden now facing you. Postcode for Satnav: LA22 9LX.

13. SIZERGH CASTLE

This National Trust property originates from a 1340 pele tower which was added to in the 15th, 16th and 18th centuries. Its sixteen acres of gardens, which date from 1740, include a splendid sunken limestone rock garden, and, amongst others, a kitchen garden and orchard.
 The castle passed into National Trust ownership in 1950 along with an almost 1,600 acre estate. The word Sizergh is of Norse derivation meaning summer pasture. Head gardener John Hawley has been there over ten years now and manages the gardens along with one other full time gardener, seasonal gardeners and volunteers. The traditional orchard contains over 50 different apple varieties with a range of tree ages from over 50 down to just ten-years-old, grown as half-standards.
 No pesticides are used but trees get an occasional organic fertilizer.

The Apples and Orchards of Cumbria

Two views of the orchards at Sizergh Castle.

64

The grass is allowed to grow so that wild flowers can flower and set seed before mowing in summer. This provides plenty of interest for the bees in the hives in the north west corner of the orchard. In addition to apples, there are also pears, quinces, cherries, plums, walnuts, fig and crab apples both in the orchard and in adjoining areas. The trees carry labels but beware of some inaccuracies.

Stepovers at Sizergh Castle.

The mostly heritage varieties were chosen for their suitability for the northern climate and include: *Allington Pippin, Keswick Codlin, Carlisle Codlin, Autumn Harvest, Duke of Devonshire, Forty Shilling, Scotch Bridget, Hawthornden, Longstart, Rankthorn, Nelson's Favourite, Galloway Pippin, Lancashire Pippin, Bloody Ploughman, Florence Bennett, Fillingham Pippin, Flower of the Town, Lord Lambourne, Hargreaves Greensweet, Fallbarrow Favourite, Gladstone, Gold Medal, Green Balsam, Harvest Festival, Hutton Square, Ladies Finger of Lancaster, Lady's Delight, Langes Perfection, Lord Suffield, Manks Codlin, Nancy Jackson, Proctors Seedling, Ribston Pippin, Sowman's Seedling, Stirling Castle, Tower of Glamis, White Melrose, Yorkshire Greening, New Bess Pool, Alderman, White Paradise, Early Victoria* (wrongly labelled *Rev. W. Wilks*), *Holstein, Charles Ross, Kidds Orange Red, James Grieve* and *Ellison's Orange*. A tree labelled *Harvest Lemon* is probably not in fact this lost variety, but more likely a *Ladies Finger of Lancaster*.

The old Fellside orchard outside the gardens in the estate was restored just after the turn of the century and was replanted in 2006 with traditional local varieties of apples, pears, plums and damsons. It is now part of the estate's garden walks. A more recent development has been the planting of 28 stepovers on M27 rootstocks around one of the kitchen garden beds in 2008. They are planted at six foot spacing between short posts at six foot intervals and trained on a single wire at two feet height. This is just enough to keep the fruit above soil splash whilst providing an attractive border to the bed without excessive shading.

The kitchen garden has a rabbit problem but so far the rabbits are more interested in the vegetables than the stepovers. Most of the varieties can be identified by copper alitags although they are hard to see. In contrast to the orchard, most of the fourteen varieties (two of each) are modern: *Greensleeves, Falstaff, Gala, Howgate Wonder, Sunrise, Red Falstaff, Sturmer Pippin, Scrumptious, Braeburn, Sunset, Fiesta, Herefordshire Russet, Epicure, Elstar*. If you are thinking about planting stepovers, the Sizergh Castle kitchen garden is a very good 'how to' guide.

Website: www.nationaltrust.org.uk/sizergh-castle. Gardens open daily 11am to 5pm from March to October inclusive (to 4pm to year-end). Entry: £5.85 (free to NT members). Directions: From M6 junction 36, go four miles west on A591 direction Kendal, then turn south on A590 direction Barrow, then first right, follow brown signs. Postcode for Satnav: LA8 8DZ. Tel: 015395 60951

14. WINDERWATH GARDENS

Winderwath Gardens may be one of the lesser known gardens in Cumbria as it is essentially a plantsman's garden rather than a tourist day out (no café gives you a clue). It is privately owned by Jane Pollock, but well established as her parents planted much of the 4½ acres which specialise in alpines and Himalayan plants along with specimen trees.

Her mother planted the orchard 40-50 years ago. The rootstocks are unknown but probably either M2 or seedling. The trees are grown as bushes, and well pruned to maintain an open centre within the classic goblet shape. Most are in good health except for two *Sunsets* dying from canker and a *Laxton's Superb* suffering a severe attack of woolly aphid. They are spaced on a five yard square which is just right for the size of the mature trees.

The orchard is grassed and mowed, with autumn crocuses attractively in flower at harvest time. The trees are unfed and unsprayed yet show only a little scab. The fifteen apples and one medlar include the following

varieties: *Bramley, Epicure, Laxton's Superb, Laxton's Fortune, Lane's Prince Albert, Lord Derby, Lord Lambourne, James Grieve, Sunset* and *Miller's Seedling*. The latter is possibly the only example of this variety in Cumbria, a second early dessert raised 1848 from Berkshire, which was grown commercially in the early and mid-20th century, but declined rapidly in the late 20th century. Most of the trees are labelled with their

identities, though you have to look hard to see the small metal labels.

Winderwath Gardens doesn't have its own website, but details can be found on: www.visitcumbria.com/pen/winderwath-gardens.htm.

This website has a couple of good aerial photos which show the orchard on the right-hand side. Open throughout the year, Monday-Friday 10am-4pm and Saturday 9am-12 noon. Location: five miles east of Penrith, half a mile east of A66, two miles north west of Temple Sowerby, 1½ miles north west of Acorn Bank. Signposted from northern end of the Temple Sowerby bypass. Postcode for Satnav: CA10 2AW. Tel: 01768 88250.

15. WORDSWORTH HOUSE AND GARDEN

This Georgian town house was the birthplace and childhood home of William Wordsworth when his father John was steward to Lord Lowther and occupied the property as a tenant. The garden and its contents (herbs, vegetables, border perennials) are set out much as it would have been during his tenancy in the late 18th century. Amanda Thackeray, the

'Perchcrow' in foreground and volunteer winter pruning of a row of Greenups Pippin the the background.

A Ribston Pippin espalier against baton mounted wooden trellis.

head gardener, has been there since 2004 during which time the garden was deluged by the great flood of November 2009.

The force of water overflowing from the adjacent River Derwent tore down several parts of the old walls which surround the half acre garden. Fertile alluvial topsoil was washed away and had to be replaced while the walls were painstakingly rebuilt. Three years later you have to look hard to see the difference, a tribute to the quality of the restoration work. The opportunity was also taken to replace grass paths with gravel which is more in keeping with the 1770s period since they didn't have lawn mowers in those days to maintain them.

The main garden contains two rows of six apple trees each, all of which are *Greenups Pippin*, the Keswick variety found in the garden of the eponymous shoemaker in the late 18th century. These were planted in the 1980s, probably on M26 rootstock and are well trained as bush trees with the classic goblet shape and open centre. This must be the largest collection of *Greenups Pippin* anywhere in the world. Quite why so many were planted is anyone's guess. Being a mid-season cooker, they don't keep in store for very long, so most are used in the kitchen and café. Nevertheless, a new cellar store is planned for some of these and the other varieties. Incidentally, the Perchcrow in the foreground of the picture on the left is so named because he fails totally to scare the birds which instead use him as a perch. He also has his own blog on facebook.

Following the reconstruction of the walled garden walls, espaliers and fans of a range of fruit types were planted in 2011. These include

Hawthornden, Ribston Pippin and *Acklam Russet* apples, *Williams Bon Chretien, Louise Bonne of Jersey* and *Catillac* pears, *Old Greengage, Mirabelle de Nancy* and *Red Magnum Bonum* plums, and *Morello* cherry. The trees are trained on wooden trellis which stands proud of the wall using batons to encourage airflow. There is also a *Brown Turkey* fig which survived the flood, and two *Portugal* quinces have been planted in one of the beds.

The adjacent kitchen garden contains more apple trees: *Keswick Codlin, Red Ladies Finger* (cider), *Fiesta* and *Golden Hornet* along with a small black bullace. This area is populated by three rare breed Scots Dumpy chickens (named Maisie, Poppy and Madam Hetty) and tame blackbirds who enjoy the windfalls.

The entire garden follows an organic regime so the trees get no feed other than compost mulch. There are no significant pests or diseases, and no mammal problems since the garden is walled and in the town. An interesting labeling feature is the use of permanent white marker on slates hung on the wall beside each espalier or fan to indicate its identity. A further feature is that this is a bumblebee friendly garden, meaning that they are actively encouraged by the wide range of cottage garden plants which provide pollen over a continuing period. Amanda explains that bumblebees rather than honeybees do the pollinating here, and she can wax lyrical on the six species of bumblebees found here, along with solitary bees and cuckoo bees. She does 'Fruits of Autumn' garden tours on Wednesdays in September which are both educational and entertaining.

Website: www.nationaltrust.org.uk/wordsworthhouse. Open daily except Friday from March to October inclusive. Admission £6.36 (National Trust members free). Location: on Main Street, Cockermouth, follow brown signs, junction of A5086 and B5292. Postcode for Satnav: CA13 9RX. Tel: 01900 820884. No car park – the nearest is Wakefield Road car park, CA13 0HG, then walk over footbridge across river.

Smaller Orchards

AT THIS JUNCTURE it is appropriate to debate what, exactly constitutes an orchard? The *Oxford Handy Dictionary* states, "an enclosure with fruit trees," conveniently declining to suggest a minimum number. *Chambers* states, "an enclosed garden of fruit trees," also avoiding mention of a minimum number. Organisations such as the National Orchard Forum and Common Ground generally claim five as the minimum number of trees to constitute an orchard, so it seems reasonable to accept that. With such a low minimum number of trees constituting an orchard, it is inevitable that many will be overlooked, so this section represents merely a sample of the smaller public orchards in Cumbria.

BROCKHOLE, owned by the Lake District National Park, is the Lake District's Visitor Centre. Situated on the east shore of Windermere it comprises the rather decaying grandeur of an attractive 19th century villa surrounded by 30 acres of a Thomas Mawson designed Arts and Crafts garden. The centre has informative displays on the Lake District's geological, social, and industrial history plus details on the events and activities taking place in the region. The café and terrace provide views over the lake with a backdrop of the Langdale Pikes. There is also an adventure playground, while a recent innovation is the Treetop Trek. Brockhole has only a handful of apple trees, most of them espaliered along south-west facing walls at the top of the kitchen garden.
Website: www.brockhole.co.uk. Gardens open daily 8.30am to 6pm. Admission free but there is a charge for car parking. The number 555 bus stops immediately outside on both sides of the road. Directions: clearly marked on A591 between Troutbeck Bridge and Ambleside. Postcode for Satnav: LA23 1LJ. Tel: 015394 46601

COWGILL COMMUNITY ORCHARD at Dent, near Sedbergh, has just seven apple trees mainly on MM106 rootstock but more planting is planned.

CROSSRIGG HOUSE is far from a smaller orchard but is placed in this category because it is rarely opened to public access. The house was built in the 1990s in the walled garden of Crossrigg Hall, which is an Edwardian gentleman's residence situated between Bolton and Cliburn, near Penrith, in the Eden Valley. The walled garden was set out during the Edwardian period with apple and pear cordon trees on the walls, and apple espaliers marking the edges of paths, plus a separate freestanding orchard for cooking apples and nut trees.

There are also some plums and Morello cherries. The cordons and espaliers, despite being about a hundred years old, are remarkably vigorous and disease free. The fertile soil and low rainfall helps, along with regular summer pruning and general good management. The 31 apple trees include nineteen different varieties while the sixteen pear trees cover ten different varieties. This garden is probably the best example of old pear growing in the county. It is open to the public only once or twice a year for charity. No website. Postcode for Satnav: CA10 3AN.

GROWING WELL is a farm-based charity about three miles south of Kendal that promotes mental health recovery by involvement in community-focused organic fruit and vegetable production. A wide range of crops are grown over ten acres of fields and polytunnels, with the produce sold through a crop share scheme. Horticultural training courses are provided and farm visits conducted for schoolchildren.

Farm manager James Smith has planted about 200 different apple varieties grown as cordons on MM106 rootstock over a three year period starting in 2008. They are planted along the boundary of the holding, representing a 200 yard long linear orchard.

Website: www.growingwell.co.uk. Directions: park at Low Sizergh Barn, beside A591 just north of roundabout with A590. Postcode for Satnav: LA8 8AE. Tel: 015395 61777

HALLGARTH COMMUNITY ORCHARD in Kendal consists of about fifteen trees growing on about a tenth of an acre plot. Most are apples including the varieties *Keswick Codlin, Yorkshire Greening, Lady's Finger of Lancaster, Orleans Reinette, Court Pendu Plat,* and more plus a few damson trees and currant bushes. From planting in the late 1990s, the tree canopies have met, calling for some hard pruning to let air and light back in and colour up the fruit. Chris Rowley has guided the project from the start and encountered the usual problems of people starting off full of enthusiasm only to drop out, so he is to be congratulated in keeping it going for so long.

Directions: Heading north out of Kendal on Windermere Road, turn right onto Garth Brow, then first right onto Hallgarth Circle then first right onto a little crescent road by a green. Halfway around the crescent there is a lane on the right between houses, walk 50 yards up the lane then turn right into Underley Road Allotments. The orchard is situated at the north west corner of the allotments.

SLOG ORCHARD in Kendal. The South Lakeland Orchard Group commenced planting a collection of local heritage apple varieties in spring 2012. The objective is to establish a reference collection of all known varieties for each of three northern counties: Cumbria, Lancashire and Yorkshire, for demonstration and educational purposes. The trees

Two views of the South Lakeland Orchard Group orchard in Kendal.

are grown as cordons, mainly on MM106 rootstocks. New varieties, particularly those containing the scab-resistant Vf gene are being evaluated along with varieties from other high-rainfall western areas to determine their suitability for the Cumbrian climate. A new rootstock, M116 is also being evaluated versus the MM106 reference, and there is also a row of pear varieties.

Directions: orchard is situated in Kendal at the southern end of Underley Road Allotments. Take Windermere Road out of Kendal, then turn right into Underley Road. After 100 yards park and walk up green lane on left.

THE NATIONAL GARDENS SCHEME Several gardens which are opened to the public under the National Gardens Scheme (NGS) have interesting collections of fruit trees. The following are examples that are often open, but consult the NGS Cumbria Gardens leaflet or www.ngs.org.uk for up-to-date details:

Ewebank Farm, Old Hutton, Kendal LA8 0NS: orchard and espaliered apples
Matson Ground, Windermere, LA23 2NH: fruit trees in walled kitchen garden
The Nook, Helton CA10 2QA: cordon, espaliered and fan-trained fruit trees
Newton Rigg College Gardens, Newton Rigg, Penrith, CA11 0AH: orchard and soft fruit
Meadow House, Garnett Bridge Road, Burneside, Kendal, LA8 9AY: vegetable and fruit garden.

Note that many of the fifteen profiled orchards also open occasionally under NGS.

Other Tree Fruit Grown in Cumbria

THIS BOOK HAS focused almost exclusively on apples, but as demonstrated in the orchards section above, other tree fruits are grown successfully in Cumbria. However, unlike apples, there is virtually no other tree fruit (except the Westmorland Damson) which originated in Cumbria.

Pears can be grown here, and despite being more difficult than apples, once established, suitable varieties of dessert, culinary and perry pears are capable of cropping well and surviving to a ripe old age.

Stone fruit do particularly well here, and none better than the famous Westmorland Damson, which has been grown in the Lyth and Winster Valleys for hundreds of years. In the 1930s, hundreds of tons of damsons were harvested annually in September, most of it going for jam-making. However, tastes change and post-war Britain lost its taste for jam so the damson industry went into steep decline.

Recently, the Westmorland Damson Association has actively promoted

Damson blossom at the Westmorland Damson Association open day.

Pear tree in blossom.

the use of damsons in artisan food production so that damsons are now a component of fruit pies, meat pies, yoghurts, ice creams, chutneys, jams, wines and liqueurs. Damson yields fluctuate considerably from year to year so the Westmorland Damson Association freezes a portion of the harvest in good years to carry over and act as a buffer to ensure continuity of supply in off-years. For more information see: www.lyth-damsons.org.uk.

Plums, being a relative of damsons also grow well in the Cumbrian climate, but are also subject to annual yield fluctuations. In both cases, they typically flower in April when the weather can be wet, windy and cold so that, unsurprisingly, the pollinating insects may not be very active.

The other major stone fruit is the cherry; that too grows well in Cumbria, but it is only worth growing for the attractiveness of its profuse pure white blossom because the crop will almost certainly be taken by the birds!

A rather marginal stone fruit here is apricot, but even this can provide fruit if

Above plums and below cherry blossom.

grown in a sheltered sunny spot, preferably fan trained against a south facing wall.

Quince can be grown as a rather spreading bushy tree, but suffers badly from quince leaf blight which causes premature defoliation from July onwards. The variety *Leskovac* is claimed to be resistant.

Medlar also grows as a spreading tree and is worth growing for its beautiful large white single flowers which continue into June, after the blossom of other fruit has fallen.

Mulberry is also a spreading tree, a slow grower but capable of doing well and yielding tasty fruit.

Above quince and below medlar flower and mulberry.

A very good reason for growing the entire range of tree fruit is to create a continuous display of blossom over a period of more than two months, starting with apricots from March followed by damsons, plums, cherries, pears, apples, quinces in sequence through to medlars in June.

Future Prospects for Apples and Orchards in Cumbria

CUMBRIA'S HIGH ANNUAL rainfall means that the pressure of diseases such as scab and canker are greater than probably anywhere else in the country. Yet the old Cumbrian varieties were originally selected because they could withstand this threat better than most other varieties. They have stood the test of time and this is why these varieties are worth preserving. The South Lakeland Orchard Group runs propagation workshops on grafting in February or March and chip budding in August; on both occasions, scion wood of Cumbrian varieties is made available to ensure the wide perpetuation of these varieties.

Orchards are more likely to survive if their produce finds a useful outlet. Many people have trees which yield far more fruit than they can use personally. Fortunately there now exist artisanal fruit juice and cider producers who will pick surplus fruit for use in their own production.

Apple varieties on display.

Lord Lambourne

Two of these cider producers have now taken the next step by planting up orchards of their own so they can ensure a consistent supply of suitable varieties such as the bittersweet and bittersharp cider varieties which cannot be found elsewhere locally.

A welcome sign is the retailing of local apple production to meet a growing demand for locally produced food. One grower sells his produce by the roadside in the Kentmere Valley. Another has taken over and renovated a former commercial orchard in Arnside and sells his produce through farm shops. Sufficient interest and demand probably exists to justify more of these initiatives.

Another positive development is the planting of community orchards. Grange-over-Sands is the local pioneer in this respect, though its history indicates the difficulties which may be encountered in terms of continuity. Other examples are Hallgarth Community Orchard in Kendal and Cowgill Community Orchard near Dent, while another is planned in Bowness. A variation on this theme is dispersed urban fruit tree planting, which has recently taken place in Windermere and Keswick. Other new orchards being planned include one at Tarnside Farm, Crosthwaite, by the charity 'Wood for Wounded', and the restoration of a centuries-old orchard at Lowther Castle.

Many schools are interested in planting a few trees as part of their gardening efforts, or even an orchard. A recent grant-aided initiative called Fruit-Full Schools selected two local schools, at Heversham and Carlisle to graft and plant their own apple trees using local heritage varieties. Queen Katherine School, Kendal, planted up an orchard of over twenty apple trees in winter 2010.

As mentioned earlier, South Lakeland Orchard Group (SLOG) and its sister organisation North Cumbrian Orchard Group (NCOG) hold annual grafting and budding workshops. These typically result in members propagating up to a hundred new trees. In addition SLOG and NCOG graft over a hundred trees which are subsequently sold at shows and apple days around the county.

The above initiatives therefore result in several hundred new apple trees planted annually in Cumbria, in addition to those purchased at nurseries, garden centres, and websites. Even discounting a few lost each year, the net annual gain is substantial.

Overall, the situation of apples in Cumbria is reasonably healthy and making steady progress. If one area could be improved it is probably the availability of locally grown apples for public sale. Interest in local produce is continually increasing, but the supply logistics need further development.

To conclude: much progress has been made in recent years, but much still remains to be done. Join SLOG or NCOG and help us do it!

Acknowledgments

I AM GRATEFUL to the owners and head gardeners of the properties described in this book and for the time so freely given to describe their orchards and management systems.

A number of sources were used for the variety descriptions, including:

The New Book of Apples by Joan Morgan and Alison Richards
The Northern Pomona edited by Linden Hawthorne
The Apple Book by Rosie Sanders
A Handbook of Hardy Fruits; Apples and Pears by E. A. Bunyard

However, most of the information on Cumbrian varieties is courtesy of Hilary Wilson and most of the information on 'Lost varieties' is courtesy of Phil Rainford. Cut apple variety illustrations are courtesy of National Fruit Collection, Brogdale.

All photographs are by the author Andy Gilchrist except *Bradley's Beauty* by Ros Taylor; *Autumn Harvest*, *Greenup's Pippin* and *Nelson's Favourite* by Jill Templeton; *Fallbarrow Favourite (1)*, *Taylor's Favourite*, *Lemon Square (2)* and *Wheaten Loaves (1)* courtesy of Bernwode Fruit Trees; *Duke of Devonshire*, *John Hugget* and *Scotch Bridget* by Jill McDonald.

For all images the copyright is the property of the photographer.

Andy Gilchrist,
October 2013

About the Author

ANDY GILCHRIST is a Westmerian who worked as an agronomist, initially in the fruit industry and subsequently in cereals. After retirement and return to Cumbria, he planted an orchard of fifty fruit trees and became Chairman and Newsletter Editor of the South Lakeland Orchard Group (SLOG).